Fresh Waters

Linda Aspen-Baxter

WEIGL PUBLISHERS INC.

Published by Weigl Publishers Inc.
350 5th Avenue, Suite 3304, PMB 6G
New York, NY 10118-0069
USA

Web site: www.weigl.com

Library of Congress Cataloging-in-Publication Data

Aspen-Baxter, Linda.
    Fresh waters / Linda Aspen-Baxter.
    p. cm. — (Biomes)
    Includes index.
    ISBN 1-59036-442-2 (hard cover : alk. paper) —
    ISBN 1-59036-443-0 (soft cover : alk. paper)
    1. Freshwater ecology—Juvenile literature. I. Title. II. Biomes (Weigl Publishers)
QH541.5.F7A87 2006   577.6—dc22   2006001033

Printed in China
1 2 3 4 5 6 7 8 9 0 10 09 08 07 06

Project Coordinator
Heather Kissock

Designers Warren Clark, Janine Vangool

Cover description: Water cascades over the rocks of Maine's Bear River.

CONTENTS

Introduction

Earth is home to millions of different **organisms**, all of which have specific survival needs. These organisms rely on their environment, or the place where they live, for their survival. All plants and animals have relationships with their environment. They interact with the environment itself, as well as the other plants and animals within the environment. This interaction creates an **ecosystem**.

Different organisms have different needs. Not every animal can survive in extreme climates. Not all plants require the same amount of water. Earth is composed of many types of environments, each of which provides organisms with the living conditions they need to survive. Organisms with similar environmental needs form communities in areas that meet these needs. These areas are called biomes. A biome can have several ecosystems.

The Rio Negro is part of South America's Paraguay River system. This system extends throughout Brazil, Paraguay, and Argentina.

Bears are drawn to the abundance of salmon in Alaska's rivers.

Fresh waters are found in lakes, ponds, rivers, and streams. **Precipitation** plays a key role in the creation and maintenance of a freshwater biome. Water falls to Earth as precipitation. When the water falls, some seeps into the soil to form groundwater. The soil absorbs some, but not all, of the water. The remaining rain becomes runoff and runs over the surface of the land, eventually developing into rivers, streams, lakes, and ponds.

These fresh waters form networks that connect to one another. Streams flow into rivers. Rivers empty into lakes. These fresh waters are home to many plants and animals who have adapted to life in and around water. All animal life needs water to survive. Fresh waters draw many species of wildlife to drink and to prey on the creatures that live there.

FASCINATING FACTS

If all the water in the world's rivers, streams, lakes, and ponds was combined, it would equal less than one percent of all the water on Earth.

About 97 percent of Earth's waters contain salt. Even the waters in a freshwater biome contain some salt, although it is much less than salt waters. It is estimated that the salt content in fresh waters is less than one percent.

Freshwater Locations

Rivers, streams, lakes, and ponds are found on every continent, except Antarctica. Land regions that were once covered by **glaciers**, such as those found in the Northern Hemisphere and in mountain regions, have the highest number of lakes and ponds. However, freshwater biomes can be found in other parts of the world as well. Most of the land in South America was never covered by glaciers, so fewer lakes are found on that continent. Instead, many streams and rivers flow downward from higher land into the mighty Amazon River.

Glaciers, such as those found in Denali National Park, cover about 10 percent of Earth's land area. They hold about 75 percent of the world's fresh water.

The Blue Nile, part of the Nile River system, plunges over Tis Abay Falls.

Located in North America, Lake Superior is one of the five lakes known as the Great Lakes. It straddles the Canada–United States border. Lake Superior covers the largest surface area of any freshwater lake in the world. Its area is 31,700 square miles (82,103 square kilometers). Victoria Lake in Africa is the second largest lake in the world. It covers 26,838 square miles (69,510 sq km). Aralb Lake in Asia covers 24,904 square miles (64,501 sq km), making it the third largest freshwater lake in the world.

Egypt's Nile River is the longest river on Earth. It is 4,145 miles (6,670 km) long. The second longest is the Amazon River in Brazil at 4,000 miles (6,404 km) in length.

FASCINATING FACTS

Lake Baikal in Siberia is the world's oldest, deepest lake. It is 25 million years old. Lake Baikal is one mile (1.6 km) deep and holds one-fifth of the fresh water on Earth.

The Amazon River drains 1,000 streams. It carries one-fifth of the world's river water and may contain as many as 5,000 species of fish.

Lake Superior's coastline extends 2,800 miles (4,506 km).

WHERE IN THE WORLD?

Freshwater biomes are found on most of the world's continents. This map shows where the world's major rivers and lakes are located. There are thousands of streams, rivers, ponds, and lakes that are not shown. Is the place where you live close to a stream, river, pond, or lake? Find the place where you live on the map. Which major rivers and lakes are closest to you?

Arctic Ocean

Asia

• Lake Baikal

• Ladoga Lake

Europe

• Lake Balkhash

Rhine River —

Volga River

Danube River —

Indus River

Euphrates
River

Tigris River

Nile River —

Yangtze River

Ganges River

Pacific Ocean

ger River —

Africa

Congo River

• Lake Victoria

• LakeTanganyika

Atlantic
Ocean

• Lake Malawi

Indian
Ocean

Australia

Murray-Darling River

Freshwater Climates

Freshwater biomes do not have their own climates. They are affected by the land biomes around them. If a river or lake is located in desert country, the climate in the area will be hot and dry. If a river runs through tropical rain forest, the climate will be hot and humid.

Freshwater biomes are affected by seasonal changes in temperate climates. Ponds and small streams may dry up in summer. In winter, ponds may freeze to the bottom. Lakes and rivers may develop a layer of ice that floats on top of the water.

Ponds are shallow, so the temperature of the water is the same from top to bottom. Lakes are larger and deeper than ponds. The water in lakes forms layers. In summer, the Sun heats the upper layer of lake water. As the water heats, it moves or circulates. The Sun's heat does not reach the deeper water, so it stays cool.

In autumn, cooler weather begins, and the top layer cools. Strong winds cause waves, which mix the layers of water. The temperature of the lake water becomes more uniform. In winter, the top layer is cooled by the cold air. It may freeze to form a layer of ice on top. This ice floats and **insulates** the water underneath from the cold air above it. The ice also stops winter winds from mixing the water. The bottom layer stays relatively warm. In spring, the air temperature warms and thaws the ice on top. The top layer warms in the Sun, and the winds again move and mix the water layers. This mixing happens twice a year in climates such as in North America. Tropical lakes can mix more or less often.

Ice floes can be found on lakes, rivers, and oceans.

FASCINATING FACTS

All the rivers in the world combined remove an average of two tons (1.8 tonnes) of rock and soil from every square mile (2.6 sq km) of land they cross every year. Rivers are earth movers.

The Lake Effect

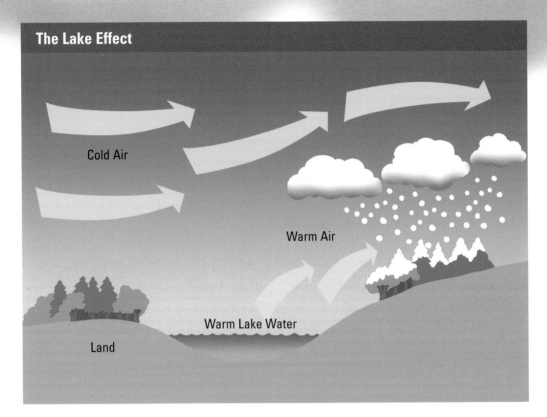

Cold Air

Warm Air

Warm Lake Water

Land

Lake-effect snowfall can leave up to 1 foot (30 centimeters) of snow within a few hours.

Depending on their size, rivers and lakes can also affect the climate. Large bodies of water absorb and hold heat from the Sun. Water warms up more slowly and cools down more slowly than air. In summer, large rivers and lakes are still cool from winter and spring, and the temperature of the water is cooler than the air. Winds off the lake or river cool the land nearby. This is called the lake effect. In winter, large bodies of water cool off more slowly than the air. This helps keep temperatures around the lake or river warmer in winter.

As well, the air over lakes and rivers contains more moisture than the air over land. This moisture, or water vapor, forms clouds and may fall as precipitation over land. Large lakes can cause heavy snowfalls on nearby land. Air that moves over a large lake becomes moist and warmer. When this air reaches land, it meets colder, drier air. This clash produces lake-effect snow.

Types of Fresh Waters

Lakes and Ponds

Lakes and ponds are inland bodies of standing water. They are found in all types of environments on all seven continents. A pond is small and shallow enough that light can reach right to the bottom. Ponds sometimes form naturally in hollows, or they can result from the building of dams, either by humans or beavers. Lakes are larger and deeper than ponds. They are fed by rivers, springs, or precipitation. Ponds are sometimes seasonal, drying up when hot weather arrives. Lakes can be hundreds, even millions, of years old.

Rivers and Streams

Rivers and streams are bodies of moving water. These bodies of water often originate at springs and lakes. Sometimes, they develop from the melting snow of glaciers. At first, they may be just a trickle of water, but they gradually become larger. Water always follows the downward slope of the land because of the pull of gravity. The flow of the water creates tiny gullies as it flows downward over the surface. These gullies meet and form bigger gullies. When water reaches a valley or has enough steady flow to create its own channel, it becomes a stream. Some streams flow only when it rains or in spring to carry meltwater. Other streams flow all year. Streams join together to form small rivers. When small rivers join a larger river, they are called tributaries. This network of streams and rivers forms a **watershed**, which drains excess water from the land.

The Lonza River runs through the Valais region of Switzerland.

The Yukon-Kuskokwim Delta in Alaska is one of the world's largest deltas.

The force of a river or stream moves **sediment** and causes **erosion**. The flow of water can be changed as a result of the new landforms that develop. Deltas, waterfalls, and rapids are just a few examples of how water charts its own course.

Deltas

A river's mouth is the place where the river empties into another body of water. As the water flows toward the mouth, it slows down. The sediment and mud that it carries sink to the bottom of the river, creating deposits. As more mud and sediment gathers, the deposits grow and become new land. This type of land formation is called a delta. Water that once flowed straight is rerouted around the delta.

Waterfalls

Waterfalls normally form in a part of the river that is at a higher elevation, such as on a hill or mountain. The water here tends to move faster than on flat land. This fast-moving water travels over both hard and soft rock. Over time, the force of the water erodes the soft rock. This erosion can cut a sharp drop in the rock, creating a waterfall.

Rapids

Much like a waterfall, rapids occur when a fast-moving river loses elevation. The drop is not as great as that of a waterfall, and the water is usually more shallow. The current is still strong, however, and the water foams around the many rocks it encounters.

FASCINATING FACTS

The world's largest delta straddles the border of India and Bangladesh. It is formed by the Ganges and Brahmaputra Rivers. The Ganges delta is approximately the same size as the entire country of Scotland.

Over time, as water drops over the edge of a waterfall, it erodes the lip of rock. Slowly the waterfall moves upstream. Niagara Falls was almost 7 miles (11 km) farther downstream 10,000 years ago.

Technology in Fresh Waters

The world's fresh waters are an important source of both water and power. Fresh waters supply drinking water and support the growth of crops throughout the world. They are also used to create hydroelectricity. Household appliances, such as refrigerators and televisions, rely on hydroelectricity for power.

Hydroelectricity is clean, renewable energy. No pollution is created to produce this type of power. Hydroelectricity is produced through the use of dams. A dam is built across a river to control the flow of water. The stored water becomes a lake or **reservoir**. Its flow is controlled by humans. When released, the water passes through turbines inside the dam. The rushing water spins the blades of these turbines to produce electricity.

The largest hydroelectric plant in the world is the Itaipú Dam on the Paraná River in South America. This power plant has 18 generating units that can produce 12,600 megawatts. They output 75 million megawatt hours per year. There is an even larger dam being constructed on the Chang-jiang (Yangtze) River of China. The Three Gorges Dam is to be finished in 2009. It is expected to output 18,200 megawatts of electricity. That is as much as 18 nuclear plants!

When completed, the Three Gorges Dam will be the largest hydroelectric dam in the world.

Fresh water is also used to **irrigate** crops. Many regions of the world are very dry. Irrigation makes it possible to grow crops in dry regions.

Different methods of irrigation are used to water crops. Drip irrigation is used with fruits and vegetables. Plastic pipes with holes in them are laid along rows of crops. Water is pumped through the pipes to water the plants.

High-pressure spray irrigation has been the standard way to irrigate crops in North America. A long tube carries water from a pump. All along the tube are triangular frames on wheels. This long arm of frames moves in a circle around the pump. The pump sends water to sprinklers all along the tube.

Low-energy spray irrigation is replacing high-pressure irrigation. A large pipe carries water out from a pump in the center. Small water sprayers hang down from this pipe. Each has a nozzle very close to the ground that sprays water gently onto the crops. More than 90 percent of the water gets to the crop.

FASCINATING FACTS

The highest dam in the world is the Nurek Dam on the Vakhsh River in Tajikistan in central Asia. It is 984 feet (300 meters) tall.

Today, about 543 million acres (220 million hectares) of the world's land is irrigated. Almost 70 percent of all the fresh water taken from rivers, lakes, reservoirs, and wells is used for irrigation.

LIFE IN FRESH WATERS

Freshwater biomes are teeming with animal and plant life. Many species of **aquatic** plants and animals live in the water. Some spend their lives on the muddy bottom, and some live near the top or on the surface. Some live in the shallows or quiet pools. Still others live on rocks in a fast-moving current. Many other plant and animal species live alongside streams, rivers, ponds, and lakes. All of these plant and animal species have adapted to living in or near fresh waters.

INVERTEBRATES

Freshwater biomes are home to many **invertebrate** species, including insects. Some insects live their whole lives in the water. Others live there for only part of their life. The **larvae** of many insects, such as mosquitoes and dragonflies, live in fresh water. When they become adults, they leave the water.

Most dragonflies remain near fresh waters following the larval stage. This is because they lay their eggs in water.

reptiles eat frogs, small fish, crayfish, insects, and mussels. Other reptiles, such as turtles, eat mostly plants. Amphibians live at least part of their lives in water. As larvae, they usually eat aquatic plants. As adults, they become carnivores and eat insects, slugs, and worms.

The Florida red-bellied turtle is found in the fresh waters of Florida and Georgia.

REPTILES AND AMPHIBIANS

Fresh waters are home to reptiles, such as snakes and lizards, as well as amphibians, such as frogs and salamanders. Reptiles and amphibians are cold-blooded. This means that they rely on the Sun to warm their bodies. When the weather and water are warm, they are more active. Many **hibernate** when the weather turns cold or become dormant if the weather gets hot and dry. Some

FISH

Fresh waters are home to two types of freshwater fish. Some are **parasites.** They attach themselves to other animals and suck their blood for food. Bony fishes, such as salmon and trout, eat plant food or other aquatic animals. They breathe with gills and use their fins for swimming. In regions where winter temperatures freeze the surface of lakes and rivers, fish survive in deeper water.

The leaves of the giant Amazon water lily can support the weight of a human adult.

BIRDS AND MAMMALS

Freshwater biomes are home to a variety of birds and mammals. Many different species of wading birds, waterfowl, and shorebirds live in freshwater biomes. Some species use riverbanks or the shores of lakes or ponds as places to nest. Mammals that spend part of their time in the water include beavers, muskrats, and otters. These animals use the rivers, streams, lakes, and ponds to find food and to drink.

The great blue heron always lives close to water. It usually nests in nearby trees or bushes.

PLANTS

Aquatic plants are grouped by where they grow in the water. Submergent plants grow under the water. Even the leaves are below the surface. Floating aquatic plants float on the water's surface. Some, such as watermeal, have no roots. Emergent plants grow partly in and partly out of the water. Their roots are under water, and their stems and leaves are at least partly out of the water. Reeds, rushes, grasses, and cattails are all emergent plants.

Freshwater Plants

Submergent Plants

Submergent plants grow entirely under water. Wild celery, coontail, and watermilfoil are just a few examples of submergent plants found in freshwater biomes. Wild celery is found in coastal freshwater inlets and waterways. It is an important food source for many waterfowl. Coontail can be found in the depths of the freshwater biome. This plant does not produce roots, so it absorbs all its nutrients from the water instead of the soil. The stems of watermilfoil can grow to 6 feet (2 m) in length. This plant normally grows in shallow waters.

Wild celery grows along the northern Pacific coast and up into Alaska.

Floating Plants

Floating plants float on the water's surface. These plants may have roots that connect them to the water's floor, or they may float freely, with no connection. The most well-known floating plant is the water lily. The flower of the water lily sits on the surface of the water, while its root is buried in the mud at the bottom of the pond. Duckweed is the smallest flowering plant. It floats on the surface in spring and summer. During this time, it produces extra starch that weighs it down. By autumn, duckweed sinks to the bottom. It uses the extra starch to stay alive during the winter. In spring, duckweed floats back to the surface again.

A thick mat of duckweed can camouflage animals, including frogs.

Emergent Plants

Emergent plants tend to grow close to the water's edge or in shallow water. Their roots are anchored in the mud of the water floor, while their stems shoot high above the water surface. Some plants can grow as much as 6 feet (2 m) out of the water. Cattails and bulrushes are the most common types of emergent plant found in the freshwater biome. Cattails are easily recognized by their flowers, which mass together to form a brown spike at the top of each plant.

Bulrushes are often a dominant plant in lakes, ponds, and slow-moving rivers.

FASCINATING FACTS

Fish and waterfowl often find their food on the leaves of coontail. The thick, bushy leaves are ideal living environments for the tiny organisms that fish and birds feed on. Coontail has more food organisms living on its leaves than any other aquatic plant.

The giant water lily is found in the Amazon basin of South America. It is the largest type of water lily in the world and can grow as large as 6 feet (2 m) in diameter.

One species of cattail, found south of Delaware in the United States, can grow up to 12 feet (3.7 m) in height.

Birds and Mammals

Wading Birds

Wading birds are just one of three bird groups that use the freshwater biome. Wading birds include herons, limpkins, and sunbitterns. These birds have long legs and wide feet for wading in shallow water. They also have long necks and bills shaped like daggers for spearing fish and frogs.

Shorebirds

Shorebirds, such as the plover and sandpiper, feed and nest along the banks and shores of lakes, rivers, and ponds. Their bills are shaped to help them get the food they need. The godwit has a long, slender upturned bill for rooting through the mud. The bill of the ruddy turnstone is curved to one side so it can turn over pebbles and shells.

The mallard defends itself by swimming or flying away.

Waterfowl

Waterfowl, such as ducks, geese, and swans, spend most of their time on the water. They have webbed feet close to the rear of their bodies to help them swim. They also have flattened bills to grab the plants on which they feed. Ducks have different feeding styles. Dabbling ducks, such as pintails and mallards, stay on the surface, dip their bills, and shake their heads. Diving ducks, such as scaups and loons, head to deeper waters. They dive deep down into the water to catch fish. Then they return to the surface.

The American golden-plover migrates from as far away as eastern Canada to the northern coast of South America—a route of about 2,500 miles (4,000 km).

Muskrats make their homes in dens alongside fresh waters.

Mammals

Many species of mammals visit streams, rivers, ponds, and lakes in search of food and water. Some mammals live in freshwater biomes and spend most of their time in the water. Mammals such as beavers, muskrats, minks, otters, platypuses, and freshwater dolphins make their homes in freshwater biomes. Many of these mammals have thick, waterproof fur. This keeps them warm and dry, even in very cold water. Some of these mammals also have webbed feet for swimming. Strong, muscular tails help propel them through the water. Even the eyes, ears, and nostrils of these mammals have adapted to life in the water. The Eurasian otter closes its nostrils and ears when it is in the water. The ear and eye openings of the platypus are closed off by folds of skin when it is submerged.

FASCINATING FACTS

Some mammals use special means to detect prey or obstacles under water. Freshwater dolphins send out sound waves that bounce off objects under water. The sound waves help them find the fish on which they prey. The platypus uses its sensitive, leathery bill to probe the mud at the bottom of the water for food.

The sunbittern is known for its colorful tail. When the bird opens its tail feathers, a pattern resembling an eye appears. The "eye" is used to scare predators.

The limpkin is often called the "crying bird." In the evening hours, the bird begins a unique shrieking cry. In the Amazon, legend says that when limpkins begin crying steadily, the river waters will stop rising.

The ruddy turnstone can fly at speeds up to 40 miles (64 km) per hour.

Amphibians, Fish, Insects, and Reptiles

Fish

Fish have perfect body shapes for life under water. Their streamlined shape helps them move through the water upstream against the current. They have gills, so they can breathe under water. They can add or release air in their swim bladders to adjust their level in the water. Some fish feed on the surface, and some prefer deep water. Trout prefer fast-moving streams. They are small and shaped like torpedoes, so they can swim upstream in search of food. Pike hide in water plants while they watch for prey. Their striped and spotted bodies camouflage them. When they see their prey, they dart out to snap it up with their sharp teeth.

Trout are a cold water fish. They need low water temperatures in order to survive.

Reptiles

All reptiles have adapted to living in water. Some, like the Southeast Asian fishing snake, can close their nostrils when they are swimming under water. Others, like the gavial of northern India, have hind legs like paddles to help them move in the water. They use their long, narrow snouts filled with sharp teeth to snap down on fish and frogs under water.

Amphibians

Frogs begin their lives as jelly-like eggs laid in the water. The eggs float just beneath the surface. When the eggs hatch, tadpoles emerge. Tadpoles have gills, so they can breathe in water. When tadpoles become frogs, they develop lungs so they can live on land. They still need water to lay their eggs and to search for insects. They have long, sticky tongues that help them capture their food. Salamanders and newts also begin their lives in the water. They hatch and live in the water as larvae. They lose their gills and live on dry land as adults. Spotted newts return to the water two to three years later to live for the rest of their lives.

Gavials are one of the longest crocodiles. They can grow up to 15 feet (4.5 m) long.

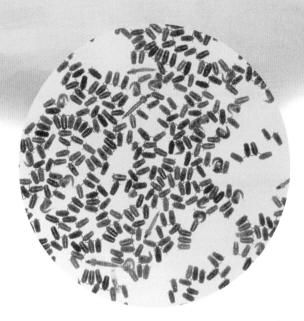

Mosquito larvae grow to between 0.5 and 0.75 inches (1 and 2 cm) long.

Insects

Insect larvae may be small, but they have ways to survive in fresh waters. Black fly larvae have suction cups to help them stick to rocks. They use head combs to catch tiny creatures from the current. Mayfly larvae have hooks on their legs for hanging onto algae-covered rocks. Mosquito larvae can tap into plant stems for air when they are under water. Fisher spiders feed on insects and tadpoles. They seem to skate across the surface of the water in search of their prey. Hairs all over their bodies spread their weight out so they can do this. The hairs on their bodies also trap air, so they can breathe under water while they hunt.

Fresh Waters in Danger

One of the most serious problems for freshwater biomes is water pollution. Pollution by factories and industrial plants poisons the water. Fertilizers, livestock waste, soil from erosion, and pesticides all enter fresh waters through runoff. Domestic sewage is dumped right into rivers and streams. When large amounts are dumped, bacteria multiply quickly. Bacteria use up the oxygen in the water, causing aquatic species to die. Streams and rivers drain large areas of land and empty into lakes and oceans. The pollution they carry travels into these bodies of water.

Water pollution threatens many forms of aquatic plant and animal life. Pollutants settle in sediment at the bottom of a lake or pond. Aquatic animals that feed on sediments eat the pollutants. Other animals prey on the bottom-feeders. The pollutants are passed up the food chain. Eventually the pollutants can end up in humans who eat contaminated fish. Water pollution poisons animals, clogs the gills of fish, and kills plant life. Polluted water also threatens water birds.

As a reaction to toxic spills, fish sometimes jump from the water and beach themselves. They die shortly thereafter.

Pollution from refineries and factories contributes to the problem of acid rain.

Acid rain results from air pollution, but it damages fresh waters. Pollutants from automobiles, power plants, and industry are released into the air. When they combine with water in the atmosphere, they form acids. These acids can be carried a long way by the wind. Eventually they fall as rain or snow, or as dry particles. Acid rain can be deadly to plant and animal life in fresh waters. Amphibians are most at risk. Their skin is thin, and the acids can pass right through it and into their bodies.

FASCINATING FACTS

In the Great Lakes, 400 toxic chemicals have been identified in the water.

Every day, humans dump two million tons (1.8 million t) of waste into the world's rivers, lakes, and streams.

WORKING IN FRESH WATERS

People who work in fresh waters play an important role in maintaining the health of these areas. They find ways to improve and protect freshwater habitats. They also learn about the vital role freshwater biomes play in the environment.

BIOLOGIST

- Duties: studies the plant and animal life that live in freshwater biomes

- Education: bachelor of science in biological science or environmental biology

- Interests: aquatic animals and plants, chemistry, environment, working outdoors, math, science, conservation

Freshwater biologists study the plants and animals that live in freshwater biomes. They also study anything that affects the natural balance in these biomes. Freshwater biologists check the health of lakes, ponds, rivers, and streams. They count populations of aquatic life at regular intervals to find out if the numbers are increasing or decreasing. They work on solutions to improve the health of freshwater biomes.

PALEOLIMNOLOGIST

- Duties: studies past freshwater life and environments

- Education: master's or doctorate degree in geology, ecology, or biology

- Interests: history of aquatic plant and animal life, conservation, environment, biology, geology

Paleolimnologists drill into lake sediment and pull out cylinders, or cores, of this sediment. They study the mud cores to learn about the past life in this freshwater environment. Paleolimnologists study the effects of pollution and acid rain on freshwater biomes. Their findings help environmental consultants and freshwater biologists.

ENVIRONMENTAL CONSULTANT

- Duties: studies environments and determines ways to protect them

- Education: bachelor's degree in environmental design or natural resource management

- Interests: environment, nature, conservation

Environment consultants study the ways in which pollution and human activity affect freshwater biomes. They look for ways to protect fresh waters from the dangers of pollution and acid rain. Environmental consultants work with biologists to help clean up polluted bodies of water.

ECO CHALLENGE

1 Which hemisphere has more lakes and ponds?

2 Why does water follow the downward slope of the land?

3 What do waterfowl and many aquatic mammals have to help them swim?

4 Which freshwater lake covers the most surface area?

5 What causes warm and cool layers of lake water to mix?

6 What do rivers dump as they enter lakes or oceans to create deltas?

7 Which insects begin their life cycle in ponds and lakes?

8 Which group of animals breathes through their moist skin?

9 Which aquatic plants grow completely under the water?

10 What is the most serious problem that endangers freshwater biomes?

WALKING ON WATER

Some insects appear to "walk" along the water surface. They can do this because of surface tension. Surface tension allows water to hold up things that are heavier and denser than water itself. A drop of water is made of smaller parts called molecules. Water molecules have bonds that hold them together. At the surface of the water, the molecules hold on to each other very tightly because there are no molecules pulling on them from the air above. As the molecules on the surface stick together, they form an invisible "skin" called surface tension. Try this experiment to see how surface tension works.

MATERIALS

- a kitchen pan
- water
- black pepper
- a clean toothpick
- liquid dish detergent

1. Fill the pan about half full with water. Sprinkle pepper onto the surface of the water. Most of it should float. Touch the tip of the clean toothpick to the surface of the water. Nothing should happen.

2. Now touch just the tip of the toothpick into the dish detergent, and return the tip of the toothpick to the surface of the water for just a moment. What happens to the pepper particles? Do they race away from the toothpick?

3. Why does this happen? The detergent weakens the surface tension of the water where the tip touches it. This allows the strong section to pull the pepper.

Water striders live on the water surface of ponds and streams.

FURTHER RESEARCH

How can I find more information about ecosystems, freshwater biomes, and animals?

- Libraries have many interesting books about ecosystems, lakes, ponds, rivers, streams, and animals.

- Science centers and aquariums are great places to learn about ecosystems, freshwater biomes, and animals.

- The Internet offers some great websites dedicated to ecosystems, freshwater biomes, and animals.

BOOKS

Hewitt, Sally. *Rivers and Ponds.* Mankato, MN: Stargazer Books, 2004.

Pielou, E. C., *Fresh Water.* Chicago, IL: University of Chicago Press, 2000.

Taylor, Barbara. *Pond and River Life.* Hauppauge, NY: Barron's Educational Books, 2000.

WEBSITES

Where can you learn more about freshwater ecosystems?

What's It Like Where You Live?
http://mbgnet.mobot.org/fresh/index.htm

Where can you learn more about lakes and the animals that live in them?

EEK! Lakes Are Great!
www.dnr.state.wi.us/org/caer/ce/eek/nature/habitat/lakes.htm

Where can you learn more about rivers and the changes they cause?

All Along a River
http://library.thinkquest.org/28022/

GLOSSARY

aquatic: living or growing in, on, or near the water

ecosystem: a community of living things sharing an environment

erosion: the process of wearing away

glaciers: masses of ice formed from snow falling over a long period of time

hibernate: to spend the winter in a sleeplike state

insulates: covers with a material that slows or stops heat, cold, or sound

invertebrate: an animal without a backbone

irrigate: to water dry land using pipes or ditches

larvae: the newly hatched, wingless form of many insects

organisms: living things

parasites: organisms that grow, feed, and are sheltered on or in a different organism; they contribute nothing to the survival of their host.

precipitation: any form of water, such as rain, snow, sleet, or hail that falls to Earth's surface

reservoir: an artificial or manmade lake

sediment: matter that settles to the bottom of a liquid

watershed: the area drained by a river or stream

INDEX